DATE DUE

Succession From Field to Forest

Willow Reed, Ph.D.

The Living World Series

ENSLOW PUBLISHERS, INC.

Bloy St. & Ramsey Ave
Box 777
Hillside, N.J. 07205
U.S.A.

P.O. Box 38
Aldershot
Hants GU12 6BP
U.K.

To Bob

Library of Congress Cataloging-in-Publication Data

Reed, Willow.
    Succession: from field to forest.
    Includes Index.
    Summary: Discusses the different kinds of plant succession,
the relationship between this process and people, and the effect
on plant communities.
    1. Plant succession—Juvenile literature. 2. Plant communities—
Juvenile literature. 3. Botany—Ecology—Juvenile literature.
[1. Plant succession. 2. Plant communities] I. Title.
QK910.R43   1991   581.5'247   90-3216

ISBN 0-89490-271-7

Printed in the United States of America

10 9 8 7 6 5 4 3 2 1

Illustration Credits:
Kathleen S. Cottrell, pp. 12, 17, 43; P.F. Narten, U.S. Geological Survey, pp. 53, 57;
National Agricultural Library, Forest Service Photo Collection, pp. 28, 51; Barry
Nehn, U.S. Fish and Wildlife Service, p. 45; R.M. Reed, pp. 24, 35, 37, 38; W.
Reed, p. 14; D.H. Richter, U.S. Geological Survey, p. 23; Sheila Saltmarsh, pp. 6,
19, 33.

Cover Photo: R.M. Reed.

Contents

1 / The Changing Landscape

The landscape around our houses and towns, along highways, and even in the countryside never seems to change very much. Many of the trees have been carefully planted and tended; the lawns, bushes, and flowers are arranged and neatly trimmed so they will look the same from year to year. We have to look at wild, untended land to find out what happens to plants when they are left alone and undisturbed. Plants change constantly in small ways, though most of the changes are hard to see. Individual plants grow each year in slightly different arrangements and patterns. Over many years, one group of plants gradually disappears and another group of plants starts to grow on the same land. Studying plants on wild land helps us understand the changes that take place, why they are important to people, and how people influence such changes in plants and plant groups.

Changing Plant Communities in Your Neighborhood

Many neighborhoods have small areas of wild or abandoned land — a vacant lot, a drainage ditch, or just the corner of a yard. You might find a patch of wild blackberry bushes next to some clumps of long-stemmed grasses and a few brightly colored wildflowers. If you look carefully underneath, you may also find small plants of other kinds, maybe even a tiny tree seedling or two.

Now imagine that nothing disturbs this little patch of land and the group of plants on it for the next several years. What will happen? Scientific studies give us some ideas. The tree seedlings may grow well, becoming large enough to shade part of the area. There will be fewer wildflowers and grasses. The blackberry bushes will be much less abundant and produce fewer berries. At the same time other bushes and trees may start to grow. Such natural changes will continue for many years as long as the area is not disturbed. One plant community gradually replaces another as conditions change.

In a city only a few kinds of plants like maples, sycamores, and lawn grass, can live where people and their activities disturb them constantly. Everything is kept neat and tidy by mowing, trimming, and

At the edge of this driveway there is a meadow covering the land disturbed when houses were built. Behind the meadow you can see some cedar trees on land that is not suitable for building and has not been disturbed for a long time.

6

using weed killers. However, wild plants start to grow on any vacant land almost immediately. Small wild parks or strips of uncultivated land along streams or ditches can also be found in cities. Often you will find oak, aspen, evergreens, or other kinds of trees growing. Underneath the larger trees there may be tree seedlings, short bushes, small broad-leaved plants, and clumps of grass. In locations such as the great plains of North America, where the native vegetation is grassland instead of forests, abundant wildflowers, bushes, and grasses grow on abandoned land or park land. The longer the land is left undisturbed the more the plant communities begin to look like wild plant communities.

Try to imagine what the land in your town looked like before any houses were built. You can make some guesses by looking at the wild plant communities growing on city land that has been vacant for many years or on the land just outside the city.

What Is Plant Succession?

What happens to plant communities? Whenever a piece of ground, and the plant community on it, is not disturbed for a period of time, a variety of different plants will start to grow. Each different kind of plant needs certain conditions to allow it to grow and reproduce. The amount of sunlight and water, daily and seasonal temperatures, and the type of soil are the most important physical conditions. Other plants also influence the growth of plants by competing for light, water, nutrients, and space. Together the physical conditions and the influences of the living organisms form the environment of the plant.

The longer the time the land is not disturbed, the more the environment changes, allowing other different kinds of plants to grow. A progression from one plant community to a new plant community with different plants growing on the same land occurs. This natural change in plant communities is called *plant succession*, and the changing communities are called *temporary communities*. Eventually a plant community develops that will stay on the same piece of land over a long period of time. The kinds of plants it contains remain the same and the environment does not change much. This type of plant community is called a *climax community*. The scientists who study plant communities as well as all other interactions among plants, animals, and the environment are called *ecologists*.

Plant Succession and the Countryside

The open countryside often seems to be filled with many different plant communities. Pastures and fields of crops, forested hillsides, ponds and streams — all appear to stay in the same place year after year. By looking more closely, you can see evidence of plant succession in these plant communities.

Along the edge of cultivated fields, small trees, bushes, and wildflowers such as bright blue chicory or goldenrod grow where the plows do not disturb the soil. Farm fields no longer being used for livestock or crops may have scattered shrubs or trees such as juniper and pine growing in them. Fields that have been undisturbed for the longest time have the greatest number of different sized trees or bushes. The plant communities in these fields are gradually changing from those with mostly grasses to those with trees and bushes.

A drive along the highway or a hike on a mountain trail can show you other changing plant communities. Recent rock slides or road cuts have no plants on them. However, as time goes by, small particles of sand and soil and plant materials such as leaves collect between the rocks. Seeds carried by wind or animals will start to grow in these cracks. Eventually the rocks will be covered by a new plant community. Next time you go for a drive, look for the oldest rock slides. Those covered with the most plants have probably been there the longest.

Plant succession may seem like a subject only for scientists, but changes in plant communities affect people in many ways. Our planet earth has a limited amount of land for the people who live on it.

Humans can no longer move to new undisturbed land when the "old" land wears out. We must learn to live with the land we have and we must understand how to use that land in the best way to support both the needs of people and the resources of the wild country. Studying which plant communities can grow and how they change is an important part of the information we need.

2/Plants and Plant Communities

Each living community is unique because of the plants and animals it contains, the environment of the community, and the interactions between organisms and environment. Because plants are usually the most obvious living part, natural communities are generally identified by the plants growing in them. *Pine forest*, *field*, *clearing*, *woods*, *thicket*, and *weed patch* are quick ways people use to identify the groups of plants they see. However, to really compare different plant communities with each other, ecologists must go further. They have to look at much more detailed information about each community and the plants growing there. It is only by comparing communities that ecologists can start to see the whole picture of different and changing plant communities.

Plant Characteristics

Describing plant communities requires recognizing the characteristics of the plants growing there. The first, most important tool ecologists have is identifying exactly which different kinds of plants, called plant species, are growing in the community. To get some idea of how to identify plants, you can try to count how many different kinds of plants you observe in a small area of a garden, vacant lot, or woods. Look at the leaves, stems, and flowers, the height and shape of the plants, and whether the same plant continues to grow from year to year as you try to decide which plants are different.

Annual plants grow from seed, produce flowers, and die in the same year. Soft-stemmed, or herbaceous, garden plants such as marigolds and zinnias are good examples. In nature, annuals are most characteristic of disturbed land, or of land that has only short periods with good conditions for plant growth such as deserts with infrequent rain showers. Perennial plants must grow for two or more years before they produce flowers and seeds. Perennials are most typical of land that has been undisturbed for a longer time and where growing conditions are good. As disturbance decreases, perennial plant species increase and annuals decrease.

Plant stems can be woody like trees and shrubs or non-woody like grasses. Almost all plants with woody stems are perennials, because it takes time for the layers of wood to grow. However, both annual and perennial plant species can have non-woody stems.

The response of plant leaves to changes in the seasons, summer to winter or rainy season to dry season, is an important characteristic of plant communities. Plants that lose all their leaves and regrow them regularly are called deciduous. Evergreen plants keep most of their leaves, or needles, year-round, making them easy to recognize during cold or dry seasons. Deciduous forests are very colorful in the fall and bare in the winter, while evergreen forests are green all year.

Arrangements of Plants in Plant Communities

All plant communities have different layers or heights of plants. The woods have tall trees and fields. Clearings or grasslands may have short bushes, short and tall grasses, and wildflowers, but no trees.

There are often plants of many different sizes in forests. The tall, mature trees form a cover or canopy layer in a forest community. A shorter layer, or understory, of young canopy trees, smaller mature

11

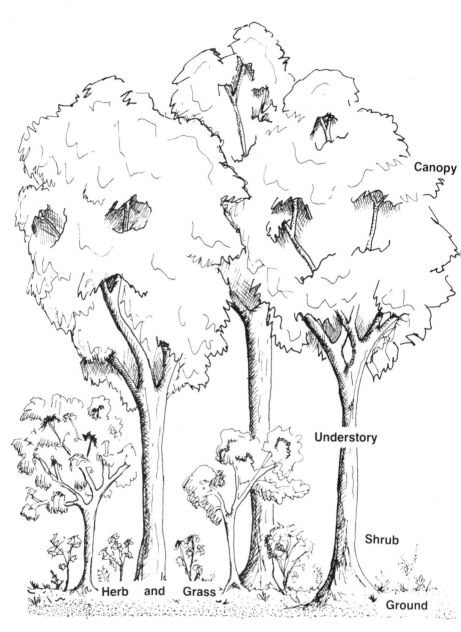

Most plant communities have plants of several different heights growing in them. This forest has four different layers above the ground.

trees, and tall shrubs is present in many forests. Tropical forests have understory layers of several different heights, while the coniferous forests of colder climates may have no definite understory layer at all.

In many forests, a shrub layer is present beneath the trees. Scattered shrubs may be the tallest layer in plant communities composed mainly of herbs and grasses. The herb layer of any plant community is made up of herbs, grasses, and often ferns. In a well-developed grassland, tall grasses and herbs may be the tallest layer.

On the surface of soil, rocks, and rotting logs, another very short layer of plants can be found. The patches of green fuzzy mosses, along with tiny herbs, grasses, and seedlings are part of the surface layer. This layer is rich in organic materials from fallen leaves and other plant parts. It is vital to the plant community because it provides a suitable place for seeds to germinate and grow.

Physical Environment of Plant Communities

Light, temperature, moisture, and soil are critical factors in every plant community. The interactions among them are very important in determining what plants will actually grow in the community. These conditions can be influenced in several ways. Climate, topography, latitude, altitude, and the type of rock or material under the plants and soil are the greatest influences. However, the plants themselves can also affect physical growing factors and can influence the future of the community.

Light is important to all plants because it provides the energy for plants to make their own food through photosynthesis. The shade and the coolness you feel when you walk through a forest are clues to what happens to the light in a plant community. The tallest plants in any community get the most intense light because there is nothing between the top leaves and the sun. The first surface the sunlight hits, whether it is plant leaves or rocks on the ground, is the hottest. The plants in the understory layers receive less light because the light is filtered through other leaves and the temperatures are cooler.

Temperature and temperature changes have a major effect on plant communities. The most critical temperature is freezing. At 0°C (32°F) plant tissues can freeze and die. Whether or not a plant can withstand temperatures at and below freezing often determines where the plant can grow. Long, regular periods of very high or very low temperatures also can affect plant growth, as can wide daily changes in temperature. Desert plants, for example, must be able to grow in locations where temperatures can change fifty degrees or more from day to night.

The availability of moisture to a plant community is another aspect of the environment. Moisture is not just the amount of water that comes in the form of precipitation (rain, snow, and fog). The quantity of water

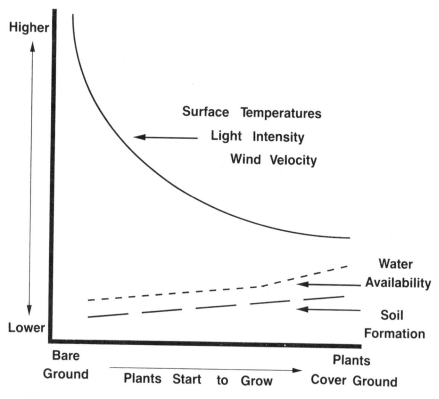

When plants start to grow on bare ground, growing conditions change. Temperature, light, and wind decrease; and water and soil increase.

that the community can actually store in the soil is critical for later plant growth. Timing of precipitation is also very important. Plants that live in areas with infrequent or seasonal rain or snow must adapt to grow and reproduce either in the dry conditions or immediately when moisture arrives.

Soil is the surface layer under the community. It is made up of rocks that have been broken down (decomposed) into many minerals important to plants such as iron, potassium, calcium, and magnesium. Mixed with these minerals are living organisms and the organic materials that come from them. Organic materials provide plant nutrients such as nitrogen, and act like a sponge to help the soil store water.

Description of Plant Communities

A detailed report of a particular plant community starts first with a physical description of the land on which the community is growing. Soil characteristics and the topography of the land which includes the altitude, the steepness of the slope, and the direction it faces are important. These are indications of the environment of the plant community. Plant communities that have many of the same species of plants frequently have very similar environmental conditions.

Second, all the different plant species in the community must be identified. The list of plant species in each community is one way some ecologists compare different plant communities. Frequently, groups of two or more plants tend to grow in the same places because their life requirements are similar. These plant groups, or plant associations, are initially identified by looking at the plant species lists in climax communities. Finding the same groups in temporary plant communities can help ecologists predict what climax community will eventually grow in that location.

One simple example of such a plant group is found in some of the undisturbed Douglas-fir forests of the Pacific Northwest in the United States. Four different shrub species (ninebark, ocean spray, snowberry,

15

and rose) make up a plant group that identifies locations where Douglas fir will grow. Whenever ecologists find these four plants together in temporary plant communities of the Pacific Northwest, they can say that eventually a Douglas-fir forest will grow there.

A third piece of information ecologists record is the number of each species in a community. Since each individual plant has some effect on changing the light conditions and using nutrients in the soil, the number of individuals of each plant species is important. This information both helps describe the plant community and gives an indication about which plants grow best. Ecologists also take into account the size of each species. Larger species like the wild rose often influence the community more than smaller species like grass plants.

A more accurate way of measuring the importance of each species to the community is to compare the actual land area covered by the leaves and stems of all the individuals of each species. The total coverage of all the individuals of one species can give us a better idea of the influence a species has on the plant community and how well it grows.

In any plant community the plant species that are able to grow and reproduce are most likely to continue to be a part of the community for a long time. The fourth part of a community description is the measurement of plant growth and reproduction. This is done most easily in forests where trees and large woody shrubs get bigger as they get older. By counting the number of small, medium, and large-sized trees of each kind, ecologists can tell which trees are growing best.

Comparison of Plant Communities

Comparing plant communities to each other is the way ecologists tell whether the communities are temporary or climax. Let's go through some of the steps an ecologist might use to compare three different evergreen forest communities that are located near each other. Most of the information will be used just to describe the forest communities. However, some of it will help tell us whether the forests we are describing are temporary communities or climax communities.

Much of northern Idaho in the United States is covered with coniferous forests that often contain just two or three different tree species. These coniferous forests represent a very valuable resource of timber and pulp wood for paper. Knowing which trees would grow best in different places helps in forest management.

Douglas fir and grand fir are the most important trees in our three different forests. The first step is to find locations or study areas where the forest is uniform or similar in appearance throughout, with mature trees and no evidence of recent disturbance by people. Sometimes this is not very easy; people have affected almost every plant community on earth in some way.

Once three study areas are found, the second step is to describe the soil and the topography. Similar plant communities have similar environments, so this information is important in helping us tell if the

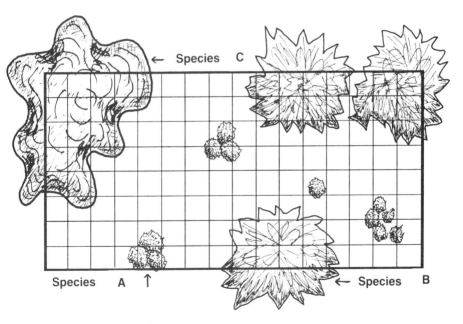

The more area a plant species covers, the more important it is to the plant community. Compare the number of plants and their total coverage in this diagram.

land could grow similar plant communities. In this case, the ecologist knows grand-fir trees grow better in cooler and moister conditions than do Douglas-fir trees.

Because the study forests will be compared with each other, the third step is measuring out a standard-sized plot or small area of land in each study area. One common size used is 50 meters by 100 meters. The specific measurements of forest plants come mostly from this plot.

The fourth step is to identify and list all the different species of plants. Our forest communities could have thirty to fifty different species of trees, shrubs, herbs, and grasses in them. Accurate identification can be a big job! By comparing the lists of plant species from different study areas, we can see if there are plant groups that might indicate that the study forests are similar now or will eventually be the same.

The fifth step is measuring the importance of the different plants in the community by estimating the coverage of all the smaller plant species and by measuring the numbers and sizes of each tree species. Size, or age class, is one way ecologists use to estimate the age of the tree. The bigger the tree, usually the older it is. The ages and numbers of different tree species tell us what trees will be in the community in the future and whether or not the community is temporary or climax.

Now let's take a closer look at the trees in the three forest areas we are studying. Forest #1 has only Douglas fir in it. They are all different sizes from young seedlings to mature trees. Since the forest is well developed and shows little sign of disturbance, we can identify Forest #1 as a climax Douglas-fir forest. It will remain very much the same over many years if nothing disturbs it.

Most of the trees in Forest #2 are grand fir with individuals of all sizes found. It seems obvious that grand fir is doing well here. However, there are a few Douglas-fir seedlings and two old trees in our plot, too. What does this mean? The two old trees tell us that sometime in the past Douglas fir grew well here. Since then, Douglas fir has not been able to grow to maturity although a few seedlings can

Even in an old forest new trees are always growing. There are many large old beech trees in this forest. However, young trees such as the small maples in the front of the picture also grow well. This climax deciduous forest will look similar many years from now, but new beech and maple trees may have replaced old trees.

get started. We conclude that this forest is a grand-fir climax forest and that Douglas fir will eventually die out completely.

Forest #3 is a lot harder to identify. There are seedlings, young, and mature grand-fir trees, but no older ones. There are seedlings, young, and old Douglas fir but no mature trees. Both species seem to be able to grow in this forest. Which one will eventually be the dominant tree? Now it becomes very important to have the list of all the plant species in the forests. By comparing the lists in the three forests, we find that the Douglas-fir forest and the grand-fir forest have different plant groups. In Forest #3 the plant group is similar to the grand-fir forest plant group. Now we can say with some certainty that Forest #3 will eventually be a grand-fir climax forest and the Douglas-fir trees will disappear after many years.

Comparing plant communities accurately takes a great deal of time and patience. However, in order to understand plant communities, how they change, and how they are affected by both human and natural disturbance, careful descriptions and studies must be done.

3/Changes in Plant Communities

Constant change is a part of all plant communities. Every season brings some alteration in plants and plant growth. Slight variations in moisture or temperature can encourage the growth and reproduction of some species of plants and not others. Occasionally, severe conditions (such as wind, rain, snow, or extreme temperatures) damage or kill plants. In a climax plant community, these short-term changes do not alter the community very much. Although conditions may not be right for every plant to reproduce every year, over a period of several years, the environment is favorable enough to allow for reproduction of all the climax plant species.

During plant succession the changes that occur in the temporary communities are much greater. Disturbing or destroying a plant community also disturbs the plant growing conditions significantly. When the disturbance stops, plant succession starts and the environment gradually begins to change back to that of the permanent climax community. What changes in the environment and in the plant communities occur and why they happen are the important questions in the study of plant succession. If we can understand such natural changes, we can start to understand how the changes humans make in plant communities will affect the future.

Environmental Changes in Plant Communities

In the time it takes for a climax plant community to develop on an area of land, some environmental factors change considerably and others do not change at all. Climate, topography, latitude and altitude, and the basic characteristics of the soil such as mineral composition and texture, do not change. This is the reason we can predict that a specific climax plant community will eventually grow on the site. However, light, temperature, available water, and soil development do change and directly affect which plants will grow at any given time.

When plants are destroyed on a piece of land, the amount of direct sunlight increases. The air temperatures at the surface of the bare ground are much higher during the day and much lower at night. With no plants to break the flow of wind, both wind and high temperatures evaporate available water much faster. As new plants start to grow, they help moderate air temperatures and lessen the drying effect of wind.

As plant communities develop on the land, the soil on which they are growing also develops and changes. Plant material is gradually added to the soil, attracting insects and microorganisms. This organic material improves the capacity of the soil to retain and store water and adds essential plant nutrients to it. By the time temporary plant communities are well established, the soil has started to develop a distinct structure with definite layers. Just as plant communities can be destroyed by disturbances such as fire or land clearing, so can soil be easily destroyed, taking many years to completely recover.

Plants do not always have exactly the same environmental conditions in the different places they live. All plants can grow with a little variation in light, water, temperature, or soil. However, as growing conditions for plants change during plant succession, the land becomes less suitable for some plant species and more suitable for others. In a climax plant community, the growing conditions no longer change to any large extent and the same plant species continue to grow and reproduce on the same land.

Plant Succession

Ecologists recognize two types of plant succession: *primary plant succession* and *secondary plant succession*. Primary succession occurs on land that has never had plants growing on it in the past. Thick volcanic ash like that from the eruption of Mount St. Helens in Washington State or hardened lava from volcanic flows on Hawaii are good examples of new land surfaces. River mud and sand deposits, rock outcrops, sand dunes, the rocks and earth scraped and deposited by glaciers, even the waste rock from mines are all areas where primary succession can occur.

Secondary succession is the type of succession most of the vegetation you see around you is undergoing. Fire, storms, floods, snow, mud

This lava flow from Hawaii's Kilauea Volcano is destroying all the vegetation it covers. When the lava cools and hardens, its surface will be an area for primary plant succession to start.

slides, and numerous human activities all damage or destroy plant communities in some way, changing the plant growing conditions. However, in secondary succession at least a few plants, plant parts, and/or seeds remain on the land to start a new cycle of plant growth.

Primary Plant Succession

The first plants that colonize the barren sites of primary plant succession must be able to live under severe environmental conditions. Nothing protects these pioneer plants from sharp temperature differences between night and day and from season to season. The new sites often have little moisture available for plant use. The wind can dry out the plant tissues or even physically destroy the plants.

On most areas of primary succession, soil gradually starts to build

Shifting sand creates the biggest problem for plants. Pioneer plants such as those in the picture have developed long and extensive root systems which stabilize the sand and allow the plants to get water. Once the sand stops shifting, other plants can start to grow.

up from plant remains and mineral particles trapped by the plants. Smaller plants are replaced by larger plants; shade increases and wind and temperature effects are lessened. Over hundreds, even thousands, of years the environment continues to change. Eventually, where there once was barren ground, a well-developed forest, grassland, or other climax community will exist.

Sand dune succession is considered primary succession because the sand is bare and unchanged by the previous activities of plants. Bare sand is a very harsh environment for plants. Blowing sand not only covers up any plants that start to grow, but it can also cut and damage the soft plant stems and leaves. Rain water runs through the sand very quickly, leaving behind little moisture for plant roots to use; the sun is very hot on the unprotected surface of the sand. With blowing sand and few plants, very little organic matter is added to the soil to provide nutrients and moisture retention. Near the edge of lakes or oceans, winds blow harder and sand dunes shift and move constantly. Farther away from the shore, the wind is less and the dunes move more slowly.

In the early part of this century a famous ecologist, H. C. Cowles, became interested in the different plant communities he saw on the sand dunes around the southern part of Lake Michigan. He noticed that the further the distance the dune was from the lake edge, the more well developed the communities on the dune were. By comparing these different plant communities, Cowles recognized a progression or succession from the bare dune to a mature climax forest. This was the first real study of how succession works. Even though many ecologists have investigated these communities since, the successional sequence of plant communities on sand dunes along southern Lake Michigan described by Cowles is still one of the best studies.

The first plants start to grow at the base of the dune on the side away from the wind, where there is more abundant soil moisture and wind protection. The pioneer plants such as Marram grass often creep up the slope of the dune by runners. As the dune becomes more stable and the conditions less harsh, a number of shrubs such as willow, chokecherry, dogwood, and grape vines invade the dune, growing above the low-growing grasses and herbs. When plants finally cover the slope, organic matter begins to enrich the soil. Trees such as linden are able to grow. The shrubs slowly die out under the increasing shade. This temporary forest community is gradually replaced by a deciduous climax forest dominated by maple and beech trees.

You may be surprised to learn that both sides of the dune do not develop the same temporary communities, though they eventually will have similar climax forests. The long gradual slope facing the wind is a much drier environment because of the wind. On these slopes, the pioneer plants are grasses and shrubs that have the ability to grow rapidly as the sand piles up around them. They are replaced by shrubby evergreens such as low junipers and bearberry. Once conditions have moderated under the cover of these shrubs, jack pine and red pine begin to grow. White pine, black oak, white oak, and eventually, the characteristic plants of the maple and beech climax forest replace the pines.

It can take hundreds of years for a climax forest to develop on a bare sand dune when there is no major disturbance. But even the temporary plant communities that develop on the dunes are important. These early communities hold the sand and help prevent erosion caused by wind and high water. Unfortunately, sand dunes are particularly susceptible to disturbance by people. Paths, digging, or other careless activity destroys the plant cover and starts dune movement and succession all over again.

Secondary Plant Succession
Disturbances caused by fire, water movement, agriculture, construction, or logging are the most common reasons for secondary succession. Exactly what plants and plant communities grow during secondary succession depends on several factors. The amount and type of disturbance, the competition among plants, changes in the variable

environmental conditions (light, soil, temperature, and moisture) and permanent environmental changes all influence secondary plant succession. The plant communities that grow immediately after disturbance are usually much simpler because fewer plant species can tolerate the severe environmental conditions. Plant communities gradually increase in complexity as succession proceeds.

One well studied example of secondary secession is the eastern foothills of the Great Smoky Mountains in North Carolina. Most of the usable land was under farm cultivation by the middle of the last century. As long as the cattle grazed or the fields were plowed, the plant communities stayed the same: low-growing grasses, herbs and hardy perennials that could resist constant cutting, trampling, and soil disturbance.

As industry developed and improving transportation systems brought food from more fertile land, people left the farms. Shrubs and tree seedlings started growing on the abandoned farm land. The fields that once grew corn or wheat or provided livestock graze returned slowly to natural vegetation. This situation gave ecologists an excellent opportunity to study how secondary succession works. Since ecologists often knew when the land was abandoned, they could tell the precise ages of different temporary communities and could record the sequence of temporary communities that succeeded one another.

Many ecologists have studied abandoned fields in North Carolina, but two in particular had a major role in describing the plant succession from abandoned field to climax oak and hickory forest. Dwight W. Billings did one of the first studies of forest development. He used study methods that could change a simple verbal description of the plants in the plant community to numerical designations for different plant characteristics. This made it much easier to compare plant communities in different areas. After Billings, Henry J. Oosting published the basic reference to old field succession in North Carolina. These studies and others help us understand how plant communities change when cultivated fields are abandoned.

Three different stages of abandoned field succession in North Carolina. Clockwise from top left: recent abandonment, pine forest, deciduous forest.

On abandoned fields, annual herbs and grasses are the first to grow well. Crabgrass, horseweed, and ragweed are common invaders during the first year after the fields are abandoned. In the second year, asters along with other herbs and crabgrass are the major plants. By the third year, broom sedge becomes established; and by the fifth year, young pine trees have started to grow. As the young pines, which need a great deal of sun, grow taller, they create shade and cooler and moister conditions underneath, helping eliminate the herbs and grasses of the first few years. The increasing shade underneath the pines allows trees that are tolerant of shade such as black gum, red maple, and dogwood to grow, while it prevents the growth of more pine seedlings. By the time the field is fifty years beyond abandonment, the pines are very tall, but many young deciduous trees are present in the understory. Maple, oak, hickory, and other slow-growing trees will eventually replace the pines. The development of an oak and hickory climax forest on the abandoned field in North Carolina can take 200 years or more.

Plant Succession and Time

One of the biggest problems in studying plant succession is that most successional changes take longer than our lifetimes. However, even though ecologists may never see what actually happens, they can make predictions about the future by comparing existing plant communities and by considering other factors.

The time plant succession takes is most obviously dependent on the types of perennial plants growing in the communities. The life cycle of a tree is often considerably longer than the life cycles of many perennial shrubs, herbs, and grasses. We would expect the whole successional history of a forest, with several different temporary forest communities, to take hundreds of years because trees live such a long time. At the same time, the succession cycle of a grassland from temporary to climax communities is much shorter because grasses and herbs produce seeds faster and are generally shorter lived.

The extent to which the land and environment have been changed by disturbance also influences succession timing. Primary succession on bare, completely new land can take a thousand years or much longer. In secondary succession, the more serious the disturbance, the longer the time it takes for climax communities to develop. One example of this is the very different effects fire can have on plant communities.

Some fires burn rapidly; they leave many plants and seeds unburned or only partially burned and do not destroy most of the organic matter of the soil. Such fires reduce the competition for light, moisture and nutrients and enrich the soil with ash from burned plants so the remaining plants and seeds grow more vigorously. Recovery of a plant community after rapid fires is relatively fast. Fires can also be very intense and destructive, burning everything from the plants to the organic matter deep in the soil. Succession of plant communities on these sites will take much longer than in lightly burned areas.

Some soil disturbances can permanently affect plant community succession. Topsoil may be removed or the soil may be compacted so much by heavy equipment that plant roots can not grow. The presence of toxic wastes can affect plant growth for a long time, as can permanent changes in the water table.

Another factor in the timing of succession is the sensitivity of the climax vegetation and its environment. Tundra plant communities, which grow on thin soil in extreme weather with a short growing season each year, take much longer to develop than do grassland communities. The tundra environment is fragile, even if the climax plants are able to grow and reproduce easily. Once there is significant destruction of the plants, the environment often deteriorates even more. For example, construction that removes plant cover in tundra areas can lead to melting of the top part of the permafrost, the permanently frozen layer of soil under the community, a critical part of the arctic environment. The time it will take for climax communities to develop is extremely variable, and if the disturbance is severe enough, sometimes the original climax community can never regrow.

4 / Climax Communities

Most of us would be very surprised if we saw a coconut palm tree growing in a forest of maple and oak. Why? Probably because we know palm trees live where it is hot and sandy, and a forest of maple and oak trees does not meet either requirement. A habitat is the place where a particular plant or animal lives. A coconut palm lives on sandy beach areas, usually near salt water, with intense sunlight, sea winds and no temperatures below freezing. In contrast, the sugar maple is found in a forest habitat with many other kinds of trees, deep soil rich in organic matter, protection from the wind, and freezing temperatures for several months each year.

Climax plant communities contain groups of plants whose habitats overlap in a particular place which has a certain set of environmental conditions. There are many different types of climax plant communities because plant species and the conditions for plant growth differ from place to place. Climate is the most general environmental influence that affects climax plant communities. However, topography, latitude, altitude, and soil are also very important.

Different Plants in Different Communities

No two plant communities have identical groups of plant species growing in them. The ability of a specific plant species to compete with other plants for light, moisture, and nutrients is the first important reason for this. For example, many crop and ornamental plants planted by people do not grow under their natural environmental conditions. To protect them from the competition of weed plants that grow much better under the same conditions, billions of dollars in weed control are spent by farmers and gardeners. In any plant community, plants that can compete best with other plants will grow best.

A near-by source of seeds is another simple, but critical reason for the presence of a particular plant species. If there are no mature, seed-producing plants close to the area, it may take years for the species to get to a suitable habitat. And as the natural habitats of plant species disappear, there is less chance of natural establishment of young plants.

Finally, the ability of a particular species to live under different conditions is important. For example, one abundant wildflower, the large-flowered trillium, is found in many types of deciduous forest. But a similar wildflower, Vasey's trillium, is only found in a few moist, deeply shaded locations because these are the only conditions under which it can grow. The presence of a particular plant species depends on its ability to adapt and grow in less than ideal growing conditions.

Characteristics of Climax Plant Communities

The first and most important characteristic of any climax plant community is that the plant species within it are able to replace themselves regularly over a long period of time. Any new, different plant species that invade, do not grow well and eventually disappear. Minor disturbances in the physical environment such as the death of a tree or a path made by animals do not affect the climax community as a whole. Usually plants from the surrounding climax community grow quickly in the disturbed area.

32

The second characteristic of a climax plant community is that the environmental conditions are predictable. Brief, unusual conditions such as extra wet or dry years do not normally affect most of the plants in a well developed climax community because the total environment is relatively stable over a much longer period of time. Temperatures and the amount of light and moisture are consistent over the years. There is a regular cycle of minerals and nutrients from the soil and from the accumulated organic matter. Unless there is a destructive disturbance such as fire, major changes in the chemical and physical balance of the environment do not occur in a climax community.

Third, the climax community is the most complex type of plant community. The community has fully developed all the plant layers

Trillium, the three-leaved plant with its upright flower, is a part of the herb and grass layer of this forest. Some species of trillium are very common because they can live in a variety of forest conditions. Others are rare because they require more specific environmental conditions to grow.

and all the habitats possible on that piece of land. Because of this, there are more different plant and animal species in a climax community than in the temporary communities that develop earlier in plant succession. Since the environment remains the same from year to year, many plants and animals found in a climax community take a longer time to grow, develop, and reproduce than those in temporary communities. The complexity and inter-dependence of living and non-living parts of a climax community make it a very stable system if nothing disturbs it. However, because the community system is so complex, any disturbance can affect the entire community much more seriously than the less complex temporary communities.

Climax plant communities are the final stage in plant succession. The importance of climax communities lies in the fact that they represent a unique picture of what interrelationships among plants, animals, and the environment can develop. Studying climax communities gives us an opportunity to understand how plant communities respond to disturbance of different kinds as well as a way of predicting the future of temporary plant communities.

Climate and Plant Communities

Climate is the combination of all the daily weather patterns over many years and the atmospheric conditions that affect weather patterns. Climate includes the temperature variations during the year, amount of sunlight, amount of moisture in the form of rain, snow, fog and humidity, and wind. Often it is not just the average temperature and precipitation that affects plant growth, but the pattern of precipitation and temperature during the year and the extreme highs and lows. For example, areas like the California coast have moderate temperatures throughout the year due to the regulating effect of ocean water. At the same time inland areas like the North American Midwest experience real extremes of summer heat and winter cold.

Climate is important in determining the general type of vegetation that can grow in a given area. Some ecologists have divided natural plant communities throughout the world into different biomes based primarily on climate. There are seven major biomes: tundra, taiga, temperate deciduous forest, tropical rain forest, chaparral, grasslands, and desert. The land areas covered by each biome are grouped together because they have similar looking vegetation and similar types of animals living in the plant communities. Although such a classification is not very specific and does not describe the many individual climax communities within each biome, it is helpful in giving us an idea of what the vegetation would look like if climate were the only major environmental influence on plant communities.

This picture of a Wyoming mountain range shows at least six different plant communities. Starting from the bottom of the picture: marsh land (with willows and shrubs on the edge), sagebrush, groves of aspen, Douglas-fir forest, and the coniferous forests of the mountain slopes.

Topography and Plant Communities

Topography influences the location of many plant communities. Mountains, highlands, plateaus, plains, rivers, and lakes are all topographic features. Along with climate, topography affects the amount of light, heat, and water plants receive. Just a little increase in the angle at which the sun hits the ground can mean a big decrease in the amount of sun energy that reaches the plant community. Slopes facing north receive the least amount of solar energy while south-facing slopes receive the most. Valleys and bottom lands in hilly areas are cooler than the surrounding areas because they are shaded by surrounding slopes and colder air settles into lower areas.

Availability of water is also affected in several ways by topography. The warmer, south-facing slopes are generally much drier with less moisture and greater evaporation of water. Permanent water sources such as lakes, streams, and the underground water table are topographic features. Because topography influences environmental conditions so much, it also influences the type of climax communities that grow on the land.

Latitude, Altitude, and Plant Communities

Latitude, the distance north or south from the equator, and altitude, the vertical distance up from the earth's surface, combine with topography to help us predict where different plant communities might grow. With changes in latitude and altitude, growing conditions for plants change. The further north or south from the equator and the higher up on a mountain you go, the colder it gets. One way we can see how climate, topography, latitude, and altitude interact to produce different climax communities is to look at the taiga forest.

At far northern latitudes, the taiga forest covers most of the land in a band hundreds of miles wide just below the tundra. However, this coniferous forest can also be found at latitudes closer to the equator under special conditions. High altitudes of taller mountains duplicate to a great extent the conditions of high latitudes. As one goes south

36

towards the equator, the climate becomes warmer and taiga forest is found in a narrow band higher and higher in the mountains. In North America, as far south as Georgia and New Mexico, the taiga forest is found on the tops of the highest mountains on slopes facing north where the climate is the coldest. On the other hand, high mountains at more northern latitudes in North America have severe enough climate to eliminate all trees. In these high mountains, taiga forests are found at the highest altitude on the south facing slopes, where they can get the maximum amount of sun energy.

Soil and Plant Communities

Soil is a fourth major environmental factor that influences plant communities. The actual mineral content of the underlying rock affects the types and availability of nutrients needed for plant growth. For

Even in the severe climate on high mountains, bushes and a few trees can grow in protected places on a warmer southern exposure.

Soil which has been undisturbed for a long time has several distinct layers. This evergreen forest soil has five layers.

example, soil that develops over limestone (calcium carbonate) will be rich in calcium and magnesium. Sometimes the soil lacks a sufficient supply of an important nutrient needed for most plant growth. The plants growing in such places must be specially adapted to grow and reproduce using less of that nutrient.

Another important soil characteristic is the size or texture of the soil particles. Water moves very quickly through soils made of large particles. These sandy soils hold very little water for plant growth. In contrast, water moves very slowly through soils made of very small particles. Clay soils, as they are called, often hold back much more water and minerals than the plants can use. In order to grow, plants must be adapted to the water conditions that the soil provides.

The depth of the soil also influences plant growth. A shallow soil cannot support plants that have deep roots. Many trees in eastern deciduous forests and many grasses in the extensive grasslands in the middle of the North American continent need deep soils. However, coniferous trees such as spruce, limber pine, and cedar grow well on shallow soil over rock, and the climax communities containing these trees are characterized by thin soil.

When people evaluate land to determine its best use, knowledge of the major environmental factors of the area and how they influence the plant communities is a valuable consideration. For productive agriculture, rich deep soil and a good water source are important. For lumber, land that will grow good timber wood reasonably fast without much extra care should be identified. Ideally, construction would be done on land that is less important for other things. Fitting the use to the land and the plant communities it supports or could support is a good reason for understanding what influences climax communities.

5/Animals in Changing Plant Communities

The presence of a particular plant community is strongly related to the external physical environment. However, all the living organisms contained within the community form part of it and influence it in some way. Plants themselves are the greatest influence, and because plants do not move, ecologists identify communities by the plants they contain. However, as you walk through your neighborhood, your eyes and ears are more likely to notice the animals.

Birds are perhaps the most obvious type of animal in any community. The least obvious, but possibly the most important animals in a community are insects and microorganisms in the soil. Larger animals such as grazing animals, which eat and trample the plants, or beaver, whose dams flood dry land, can sometimes be the cause of major changes in plant communities. Every animal, large or small, plays a part in the whole community.

Plant and Animal Interactions

Interactions between plants and animals are important for both. Plants and plant communities provide food and shelter for every kind of animal. Insects transfer pollen from one plant to another, essential for

40

seed production in many plants. Many animals and birds eat edible seeds and fruits and spread them to other locations. The underground activities of insects and other ground-dwelling animals help loosen the soil, mix plant materials and other organic matter with the soil, and allow air to circulate around the plant roots.

Occasionally animals affect the community so much that the plant community is changed. When the natural distribution of animal species is altered or the population increases or decreases dramatically, plants are often most directly affected because they can not immediately react or respond to changes. Locust swarms that eat all the vegetation in their path have been a devastating problem for humans for thousands of years. And since people have been able to travel easily from one part of the world to another, they have frequently introduced animals and insects to new lands where they prove to be destructive.

Plant Communities and Animal Habitat

The needs of animals for food, shelter, and reproduction determine where they live. You already know where to find many common animals. You would expect to see a woodchuck (ground hog or marmot) in an open field, along the edge of the road or in a garden. Squirrels and woodpeckers are found in trees, and many kinds of mice and ground-dwelling birds such as quail are likely to be found in meadows. There always seem to be more butterflies in a field than in a forest, and dragonflies are most commonly found near water.

Many animals are found in general plant community types such as forest, grassland, desert, tundra, or wetlands. These animals have no requirements for specific plant species. Large herds of grazing animals such as the North American buffalo used to live on the great expanses of grassland. The zebra, wildebeests and the wide variety of other African grazing animals are at home in a tropical grassland habitat with scattered trees. The black bear, found in both coniferous and deciduous forests, needs a habitat with trees and a variety of foods. Deer live in locations where they can browse on shrubs and herbs and where they can move quickly from open areas to wooded areas for protection.

Other animals use different plant communities at various times during the day or year. For example, wapiti or elk live in the high mountains of western North America in the summer but spend the winter in valleys where the weather is less severe and food is more available. The variety of different plant communities available in a limited area can be important to some animals. The ruffed grouse requires newly created forest opening with herbs and low shrubs for summer food, dense stands of young trees for nesting cover, and mature forest for winter food.

Some animals can only live in a very specific habitat. The pica is a small, round-eared rodent found only on rocky slopes in the spruce-fir forest in western North America. Occasionally animals depend on one or a few plant species that can only be found in particular plant

These deer are grazing in a pasture next to a forest community for protection.

communities. The giant pandas of China are totally dependent on bamboo for food, and the koalas of Australia are dependent solely on eucalyptus trees. These animals are threatened with extinction because the plant communities that provide their food are being destroyed. The more limited the habitat requirements of animal species, the less likely they are to survive if the plant communities in which they live are disturbed or destroyed by humans or by nature.

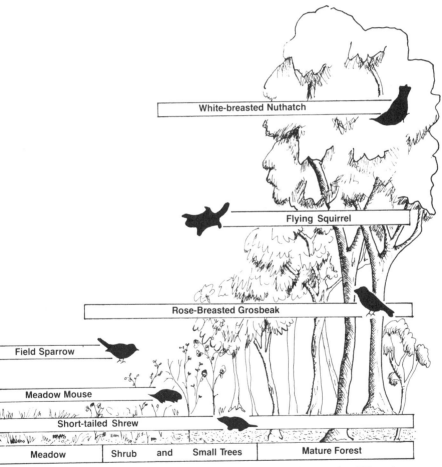

Different species of birds and animals live in different types of vegetation. When plant communities change, sometimes birds and animals change, too.

Plant Communities and Diversity of Plants and Animals

The diversity of plant and animal species in plant communities varies a great deal. As temporary communities change to climax communities during succession, the number of plant and animal species increases. Diversity also varies in different climax communities. The tundra of the arctic has relatively few kinds of plants and animals because of its severe environment, but the greatest possible number of each species is reached in climax tundra communities. Further south, the climate gets less severe and species diversity increases. Climax deciduous forests have a greater variety of species than the climax coniferous forests of colder climates.

The number of plant layers in the plant community is very important in determining both animal number and diversity. In general, the more different plant heights there are in a community, the more diverse the animal life is. Grasslands with only a little variation in plant height support large numbers of a few animal species. Tropical forests with many different plant layers have the greatest number of animal species, many of which live in specific layers. If plant layers are missing, the animals that require the special habitat of that layer cannot survive.

As communities are changed by people, many species are either reduced in number or become extinct. Since we seldom are even aware of the missing organisms, why is this important? From a practical point of view numerous valuable uses—from foods to life-saving medicines—continue to be discovered for some little known plants growing in the protection of natural communities. Without the refuge of undisturbed plant communities, such living resources, many still undiscovered, may be lost forever. From a broader point of view, destroying irreplaceable pieces of the complex plant and animal community changes our world permanently, leaving less to learn and less to share.

Plant Succession and Animals

How does this relate to succession? Both temporary and climax plant communities help determine the animal life we see around us. While many animals are only found in undisturbed climax communities, many others live best in temporary communities or a mixture of communities. Often the boundary between different plant communities such as the edge of a cultivated field or the border between woods and grassland is a refuge for a great variety of plant and animal species. These areas combine characteristics of two plant communities and are less subject to continuing disturbance.

Plant succession can be directly affected by animals. The pine bark beetle can destroy whole forests of pine trees, delaying the development of climax plant communities. Beaver dams flood large areas of land, providing food and protection for the beavers, but killing much of the surrounding plant community. When the water finally drains off, a new successional sequence is started. In the Rocky Mountains mule deer and elk can actually prevent the growth of trees and shrubs in newly made forest openings by eating the seedlings of these plants for winter food.

Plant Communities, Animals, and Humans

Because plants are dependent to a large extent on animal activities, anything that affects animal numbers can affect the plant community. Destruction of pollinating insects can affect the reproduction of many plant species, and destruction of insect-eating birds can result in much greater numbers of harmful plant-eating insects. Poisoning the soil with insecticides or toxic wastes kills beneficial soil organisms that are needed to ensure the natural community changes. Killing off animal predators such as mountain lions, wolves, and hawks can allow deer and rodents such as mice and rabbits to increase greatly. As the numbers of these animals increase, their need for food increases, and the natural changes in plant communities are slowed because more plants are being eaten.

45

The presence of certain animals in different plant communities depends on the availability of the right habitat. The general vegetation type, the successional stage of the plant community, the presence of disturbance, and the overall complexity of the vegetation all are important. As the world continues to disturb and destroy natural plant communities, the destruction extends to the animals. Each part of a natural community is important to the whole. Any changes, whether natural or human, can affect animal life as well as plant life.

6 / Human Activity and Plant Succession

Human activities have affected natural plant communities and plant succession throughout time and around the world. Only remnants of original climax plant communities are left in countries like Great Britain and the Netherlands where human activities have been intense over hundreds of years. In places such as Africa and India, human populations are growing so rapidly that all available land is being cleared for growing food and providing firewood.

Some human effects are direct. Use of land for farming, tree harvesting, constructing buildings and roads, mining, and building dams are examples of common disturbances that remove plant communities from the landscape. Other effects are indirect, often resulting in even more widespread damage to plant communities. Pollution of water, soil, air, fires, and the accidental introduction of diseases, destructive insects, and other animals, all either destroy climax communities or interfere with the normal succession of plant communities. One of the most serious questions of the future is how humans can work with the lessons of natural communities to provide for agricultural products and other human needs without destroying irreplaceable natural plant, animal, and ecological resources.

Agriculture

Natural, undisturbed plant communities almost always produce more total plant material than artificially maintained agricultural fields do. In an undisturbed community the many different plant species each use slightly different nutrients from the soil so plant nutrient stores are rarely depleted. The fact that the natural community contains many plant species means it is more resistant to complete destruction by many insects and diseases. Many of these generally attack just one or two species. However, for people the drawback of most natural, undisturbed communities is that little of the plant material can be harvested for human use. As human populations increase, people try to use the existing land to its maximum extent for food and materials. As a result, agriculture is a common, necessary disturbance of natural plant communities.

Growing single crops like wheat, corn, or even trees is an efficient way of producing a product for human use. However, because all the plants of a single species need exactly the same nutrients, additional nutrients must be artificially added to the soil. Constant cultivation of the soil also reduces the plant nutrients and important soil organic materials because it leaves the land susceptible to erosion, the wind or water removal of soil. Land planted with a single crop is much more likely to be damaged by insects, disease, or natural events like wind and hail. Weed plants invade easily because they are well adapted to disturbed conditions, especially when other competing native plants have been removed.

Most farmers regulate the growing environment constantly with chemical fertilizers and weed and insect killers. Selective breeding and genetic manipulation are used to develop high-yielding crop plants. As a result, the amount of usable product produced by a single-crop field is far larger, in the short term, than in natural communities. However, intensive management practices destroy soil, reduce fertility, and leave toxic residues, affecting the future of the land and the plant communities it could support.

The pressure for more food means every available piece of land is often put into cultivation. The strips of land separating fields in developed countries serve as refuges for a variety of wild animal and plant species and provide seed sources for abandoned land. When these are destroyed to put more land into production, a valuable plant and animal resource for reestablishing natural communities is eliminated. In grain producing, grassland areas of the world, natural vegetation strips are very important as wind breaks. Strong winds can move tons of fertile top soil on the flat land without taller vegetation to disrupt the air flow.

Production of animal products by humans has similar problems. Domesticated animals, many of which eat the same plants, are often confined to the same pasture year-round. Grazing too many animals for too long a time on the same land destroys both the plants and the soil surface and structure. Wild grazing animals living under natural conditions are normally not so destructive. In North America the enormous buffalo herds used to migrate long distances during the year, allowing the grassland time to recover from grazing. On the African plains, many different kinds of grazing animals each eat slightly different food plants and move with the seasons. Animals growing under natural conditions produce up to fifteen times as much meat per unit of land than do cattle.

There are a number of agricultural practices that do use lessons learned from natural communities. Yearly rotation of crops and cover crops, which are plowed under instead of harvested, can help restore nutrients and organic matter to the soil. Some farmers are experimenting with reduced or limited cultivation of the soil. Principles of organic farming are being considered for some commercial agriculture. This involves building up the soil and health of the plants rather then relying on artificial fertilizers and toxic chemicals. Domesticated animals can be shifted from pasture to pasture during the year. A variety of animals or mixed plant cropping can yield more production from the same amount of land. Using such techniques not only helps agriculture now, but it also helps enrich and preserve resources for the future.

Forest Management

Forests provide an important and rapidly disappearing resource all over the world. Even in this age of technology we depend on wood for fuel, paper, and construction as well as tree products such as rubber, turpentine, and food. Knowing the history, present condition, and probable future of a forest means better decisions can be made about how it should be managed. Plant succession studies have an important role in this information.

Three forest management techniques reflect our understanding of the way forests grow and change. Severely burned or heavily logged forest areas can be reforested by planting trees that can survive in the open sunlight. Leaving areas of uncut forest between areas of cleared forest provides a natural seed source for plants on the cut-over land. Selective cutting of only the larger timber trees reduces the amount of disturbance, leaving many smaller trees to continue to grow as seed sources and future timber.

Maps of forest communities actually growing on the land are forest management tools. By using forest succession studies, some of the following questions can be answered. Are the forests likely to change in the future? What type of forest will eventually grow best on the area as a climax forest? How fast are the different tree species growing? Where are different tree species growing best? What diseases or harmful insects affect the area? How sensitive is the forest to disturbance?

Maps and descriptions of plant communities have broader uses too. They can help in making informed decisions about the many possible uses of forests, grasslands, or other areas. Commercial uses, recreational uses or protected natural areas are all possible choices.

Construction Projects

The most common plant community disturbance in populated areas is construction for buildings and highways. Conditions for plant growth are drastically changed. Plants are killed and seed sources are eliminated. The upper layer of soil containing most of the organic matter and many nutrients is stripped away, and the remaining soil is often mixed with waste materials on the site. Heavy equipment on the site compacts the soil, making it difficult for plant roots to grow.

Once construction is finished, the unused land is sometimes partially restored by covering it with new soil and planting some sort of plant cover. Often, however, nothing is planted and wild vegetation takes over. The first plants to grow are mostly annuals that can tolerate the severe conditions. As the growing conditions change, temporary plant communities start to develop. Which plant communities will

Alternating blocks of completely cut and uncut forest provide natural seed sources for new forest communities on cleared land.

grow first and the length of time it will take for climax plant communities to grow depends on how extensively the land was disturbed and whether disturbance continues.

Mining

Mining activities can both directly affect the land through physical destruction of the land and indirectly through toxic chemicals. In Sudbury, Ontario in Canada (and around many other mining operations throughout the world) the processing of ore releases large amounts of a chemical called sulfur dioxide into the atmosphere. When combined with water in the atmosphere this gas forms sulfuric acid, and becomes acid rain which kills surrounding plant communities. Destruction of these plant communities is made even worse by two other factors: the shallow soil of many mining areas and the fire hazard of dead vegetation. When the plants burn, the thin soil surface has no protection from water and wind erosion; the soil covering the bedrock is lost. The bare rock surface is a difficult environment for plants to establish again, even if the periodic chemical releases are stopped.

People can reduce or even reverse such destruction. In many cases laws and financial rewards have helped. The toxic fumes from the Sudbury ore processing have been reduced by law. Their levels are now five percent of what they were just fifteen years ago, and the mining companies are making money from the waste products which have been so destructive to the environment.

Air Pollution

Air pollution is another way human activities can affect plant succession. Acid rain results from the release of sulfur dioxide gas from the burning of fossil fuels (oil, gas, and coal) for energy and is recognized as a major problem for plant communities and in the long run for people. The most immediate effects are seen in lakes and streams where water gradually becomes more acid. In some areas fish and other water life have already been seriously affected, even destroyed, by high acidity.

Plant communities are also starting to be affected. Measurements of the growth of coniferous trees in Germany and in the Appalachian Mountains of North America show that growth has slowed down over the last few years because of acid rain. How will this affect the future of forests and the forest products we depend upon? We do not have the answers yet.

Disease, Insects, and Plant Succession

Diseases and insects that affect important plant species can modify plant communities. Ecologists do not always know what the long-range effects will be, but looking at an example gives some clues. The

These abandoned coal mine wastes in Utah are over 30 years old. The background hills are covered with natural piñon pine/juniper plant communities, but few plants grow on the waste piles because of high surface temperatures and low water availability.

American chestnut tree used to be an extremely valuable timber and nut tree, comprising as much as twenty-five percent of the eastern deciduous forest as an important member of the climax plant communities. At the beginning of the twentieth century, a tree fungus called chestnut blight was accidentally introduced into North America when the Chinese chestnut tree was imported. This disease attacked and killed the American chestnut, spreading gradually south and west through the forests. Today only a few scattered individuals and groves out of the entire population of this majestic tree are still alive.

As a result of the chestnut blight, forest communities in the eastern North America were affected in two different ways. Throughout the area, forest openings were created wherever the chestnuts died. These have undergone secondary succession, often repopulated by climax species of the surrounding forest. Of more importance is the fact that the chestnut is no longer part of the climax plant community. Other trees such as sugar maple, oak, hickory, and tulip poplar trees have increased in abundance in areas where the chestnut was once present. Succession in the forest communities, the composition of the forest, and the use of the chestnut products such as nuts and wood by humans and animals have all been changed in a major way by the chestnut blight.

Other major forest trees are currently seriously affected by disease and insects. Dutch elm disease also introduced by humans, continues to kill elm trees across eastern North America. Infestations of the destructive gypsy moth that eat most forest vegetation, the southern pine bark beetle and the spruce bud worm are of concern to foresters and ecologists alike. They threaten the economic products that natural and managed forest communities provide. It is important to study the conditions that promote the spread of plant diseases and harmful insects both to prevent plant destruction and to understand what their presence does to the future of the plant communities.

Human Activities and Permanent Changes in Plant Succession

Though we like to think that many disturbances of plant communities are reversible, even simple disturbance can cause permanent change. Old covered wagon routes across western North America are still visible more than one hundred years after they were used. Where the soil has been compacted, the growing conditions have been changed. The trail ruts, not used for many years, still grow smaller plants and some different plant species. Will they ever look the same as the rest of the prairie grassland? No one really knows.

Some human disturbance is much more serious. Perhaps the most widespread permanent damage humans are causing now is in the tropical rain forest. The rain forest may seem like a far-away world to most of us, but its importance is tremendous. The complex plant and animal life is a resource for new products and uses that we have just barely started to appreciate. Of even more critical importance is the chemical balance of the earth's atmosphere that the vast areas of tropical forest and other vegetation help control. The use of fossil fuels such as coal, oil, and gas is releasing large amounts of carbon dioxide into the atmosphere. But tropical forests, which use great quantities of carbon dioxide, are being destroyed at the rate of an area the size of the state of Connecticut every year. There are real concerns that the increase of carbon dioxide is contributing to the long-term warming of the whole earth. This warming is called the greenhouse effect.

Typically, agriculture in tropical rain forest consists of clearing patches of forest and burning all the plant material to increase the fertility of the soil for crops. This is called "slash and burn" agriculture. Soil fertility, which is maintained in the actively growing rain forest, decreases rapidly under cultivation because of warm temperatures and abundant rainfall. Erosion is greatly increased, weeds and insects become more troublesome, and frequently the land is abandoned after a few years. Sometimes the opportunistic plants from the surrounding areas, large leaved herbs and rapidly growing, soft wood trees, quickly cover the abandoned land. However, often there is another even more serious problem—permanent changes to the soil.

When land in tropical areas of the world is deforested to grow crops and the bare ground is exposed to the sun for as little as three or four years, the soil often hardens so much that the native rain forest community can never grow back. Fires often become frequent after deforestation. The vegetation permanently changes from the productive and complex tropical rain forest to much simpler and less fertile grass and shrub communities with widely scattered trees.

Restoring Plant Communities
Damaged by Human Activity

The direct disturbance of plant communities by human activity does two things to the normal course of plant succession. First, it can simply start the successional sequence all over again. Second, it can actually affect the environment so much that it will take much longer or in some cases never again be able to grow the original climax community. It is hard to predict how permanent the effects of many direct human interactions with plant communities will be. However, even the awareness that human activities could permanently damage the land and the plant communities is an important step.

Sometimes after land is disturbed, undesirable plants grow vigorously, preventing native or more useful plants from growing. To prevent this, knowledge of plant succession can be used. On the dry grassland east of the Rocky Mountains, abandoned fields are often subject to severe erosion. This land may never to return to natural vegetation. Weed plants with little value for grazing animals grow too well under these conditions. However, when cultivation is stopped, some farmers promptly plant the same grasses found in the local, natural grassland communities. These native grasses do very well. Within three years the grassland is as good for animal grazing as land that has never been disturbed.

Knowing what species of plants will grow best is only part of the answer; the second part is identifying the environmental conditions needed to allow the plants to grow. The amount of light, moisture, and necessary plant nutrients in the soil are important in the success of reestablishing plant communities. Surface mining activities destroy the land by stripping all the vegetation and ore-rich rock from the land. The raw rock that remains often contains compounds which kill plants; there may not be enough water or plant nutrients for years. In order to reestablish plant communities on such areas, not only must the right plants be planted but fertilization and watering of the plantings must be continued for many years.

Indirect effects of humans such as diseases, insects, and pollution leave us with questions to answer about the future. Following is a list

A natural diversity of vegetation has been reestablished on this area of mining waste. There are transplanted aspen tree clumps, rock piles, and plowed furrows with planted grasses along the natural contours of the slopes.

of important questions still unanswered:

- What were plant communities like thousands of years ago, and how have they changed?
- How do different types and severity of disturbances, and past history and composition of near-by plant communities affect secondary succession?
- How do ideas about climax and successional plant communities apply to the tropics?
- What can people learn about tropical forests when only small areas of climax tropical forest still exist?
- How will the current destruction of large areas of tropical forest affect plant communities and the earth's environment in the future?
- How will acid rain affect plant communities and human use of land in future years?
- How can humans maintain diversity of plant and animal species in natural habitats?

In many cases we do not know exactly what the effects will be. The sensitivity of the plants, the kind of disturbance, the intensity of the disturbance, and the continuing presence of the disturbance all need to be considered. Our knowledge of what could happen if changes affect plant communities must be based on understanding what happens during natural changes in plant communities during plant succession.

Glossary

acid rain—sulfur dioxide gas combined with moisture in the atmosphere increases the acidity of rain and snow, making it more harmful to plant growth.

adapt—to adjust to the conditions.

age classes—different sizes of a plant species present on a piece of land.

altitude—the vertical distance up from the earth's surface.

annual plant—a plant that grows, produces seeds, and dies in the same year.

biome—large, easily recognizable, very general units of vegetation with similar types of plants and animals.

canopy—tallest vegetation in a community, usually referring to trees.

climate—interaction of all the long term atmospheric factors that affect an area of land over many years.

climax community—the final stable plant community in plant succession.

community—plants and animals interacting and living together.

competition—result of a common demand of two or more organisms for environmental resources (light, water, nutrients, space) in short supply.

coniferous—trees with needle-like leaves usually evergreen, with seeds in cones.

coverage—the actual amount of land the above ground plant parts cover.

deciduous—plants that lose all their leaves in the winter or in the dry season.

diversity—variety of different species of organisms.

ecologist—a scientist who studies the interactions between living organisms and their environment.

environment—physical and biological conditions and surroundings of a plant or animal.

erosion—the removal of soil or sand by wind or water.

evergreen—plants that have at least some leaves or needles all year-round.

greenhouse effect—increased carbon dioxide in the atmosphere that can increase average temperature of all parts of the earth.

grass—non-woody annual or perennial plants with long, slender leaves and inconspicuous flowers.

habitat—the place, identified by environmental characteristics, where a plant or animal lives and reproduces.

herb—a non-woody plant, perennial, biannual or annual, whose above ground parts are relatively short-lived.

identification—naming of specific plant or animal species.

invader—plants or animals that enter and grow in an existing plant community.

latitude—the distance north or south from the equator, measured in degrees.

life cycle—cycle of life in any organism from beginning through development and reproduction to death.

organic—any material from living or once living organisms.

perennial plant—a plant that must grow for two or more years before it produces seeds.

permafrost—permanently frozen layer of soil in arctic climates.

photosynthesis—the chemical process by which plants convert sunlight into food energy.

pioneer plants—the first plants to live on an area which has never had plants growing on it.

plant community—groups of plants interacting and living together in a particular location.

plant groups—two or more plant species that are found regularly growing together in both temporary and climax plant communities.

plant succession—natural progression of plant communities over time, leading to a stable plant community that does not change significantly unless disturbed.

plot—regular sized area of land, used for measurements and comparisons.

primary succession—plant succession that begins on new or bare land surfaces that has no plants, plant parts, or seeds.

secondary succession—plant succession that begins on land that has a disturbed plant community and some viable plant seeds or parts.

shrub—perennial woody plant, less than 9 feet (3 meters) and often having several stems.

soil—the top of the earth's crust made up of minerals from rocks mixed with living organisms and products of their decay.

species—a group of interbreeding plants or animals.

temporary communities—plant communities that develop in plant succession before the establishment of a final stable, climax plant community.

tree—perennial woody plants, growing at least 16 feet (5 meters) tall and usually having a single trunk.

topography—the physical characteristics of a land area.

understory—layers of vegetation under canopy, usually referring to trees.

vegetation—general definition of plant life growing together.

water table—the level at which underground rock or soil is saturated with water.

weather—condition and characteristics of the atmosphere at a given time.

weed—a plant that is out of place and usually has the ability to grow in a wide variety of environmental conditions.

Further Reading

Facklam, Howard and Margery. *Plants: Extinction or Survival?* Hillside, N.J.: Enslow Publishers, Inc., 1990.

Lambert, David. *Grasslands.* Englewood Cliffs, N.J.: Silver Burdett Press, 1988.

Lye, Keith. *Deserts.* Englewood Cliffs, N.J.: Silver Burdett Press, 1987.

Pringle, Laurence. *Restoring Our Earth.* Hillside, N.J.: Enslow Publishers, Inc., 1986.

Scott, Jane. *Botany in the Field: An Introduction to Plant Communities for the Amateur Naturalist.* Englewood Cliffs, N.J.: Prentice-Hall, 1983.

Schoonmaker, Peter K. *The Living Forest.* Hillside, N.J.:Enslow Publishers, Inc., 1990.

Zim, Herbert S. and Alexander C. Martin. *Trees: A Guide to Familiar American Trees.* New York: Golden Press, 1956.

INDEX

North America, 7, 34, 37, 49, 54
nutrients, 15, 22, 30, 33, 48

O
Oosting, H.J., 27
organic matter, 15, 22, 25, 26

P
perennial plants, 11, 27, 29
photosynthesis, 13
plant groups, 15, 18, 20
plant layers, 11-13, 33, 44
pollution, 47, 52, 57
precipitation, 14, 15, 34

R
rain forest, 35, 55, 56
rocks, 13, 23, 37, 52, 57

S
sand dunes, 25-26
seeds, 13, 24, 29, 32, 41
shade, 13, 25, 29, 32, 36
shrub layer, 11-13
shrubs, 11, 26, 27
slopes (hillsides), 15, 26, 36, 37
soil, 15, 30, 37-39, 56
species, 10, 17, 42, 43, 44
succession, 44, 47, 56
    forest succession, 50
    plant succession, 7-8, 23-30, 50,
        55
    primary succession, 23-25
    secondary succession, 23-24,
        26-29, 54
sun, 13, 25, 36, 56
sunlight, 13, 22, 34, 50

T
taiga, 35, 36-37
temperature, 14, 22, 24, 34
timber, 17, 39, 50, 54

topography, 13, 15, 22, 36
toxic materials, 30, 45, 49, 50, 52
trees, 11, 53, 54
tundra, 30, 35, 36, 41, 42

U
understory, 11-13, 29

V
vegetation, 35, 46, 52, 54, 55, 56, 57

W
waste materials, 30, 45
water, 14, 22, 36, 39, 52
water table, 30, 36
weather, 30, 34, 42
weeds, 55
wind, 22, 24-26, 34, 49